David S. Fielker and Jos

SQUARES

CAMBRIDGE UNIVERSITY PRESS

Published by the Syndics of the Cambridge University Press
Bentley House, 200 Euston Road, London NW1 2DB
American Branch: 32 East 57th Street, New York, N.Y. 10022

© Cambridge University Press 1974

ISBN: 0 521 20025 3

First published 1974

Photoset and printed in Malta by St Paul's Press Ltd

ACKNOWLEDGEMENTS

Thanks are due to the Keystone Press Agency for permission to include the picture of the game of chess on p. 3. The picture shows the Yugoslav Grandmaster Svetozar Gligoric playing chess with a pupil of a school in Wallasey, England.

Chessboard squares

How many squares are there in a chess or draught board?

Did you say '64'? Yes, there *are* 64 at first glance but suppose you count larger squares like these:

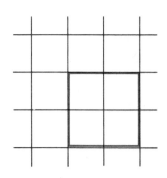
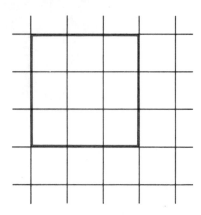

There are other sizes too. See if you can find the total number of squares of all sizes. You will be able to check your total after some work later in this book.

Squares on a nailboard

Have you got any nailboards, or geoboards as they are often called, in your classroom? You will need at least a 9-pin, a 16-pin and a 25-pin board, as shown in the diagram, and some elastic bands.

If necessary they are easy to make using a wooden base and small nails. Squared paper over the wood makes a useful guide for positioning the nails and can be ripped away when they have been hammered in. If you have no boards and cannot make any, then you can work at this section using paper, with dots instead of nails and pencil lines instead of bands; again squared paper is useful.

Use elastic bands to see how many *different* sizes of square you can make on the 9-pin board.

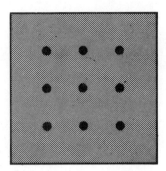

After the chessboard problem you probably found at least two sizes. Did you find *three* sizes? If not try again. The diagram opposite might give you a clue.

Now make squares of different sizes on the 16-pin board.

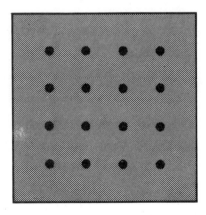

Can you find all *five* sizes?

Most people seem to find the one marked 'd' in this diagram the most difficult.

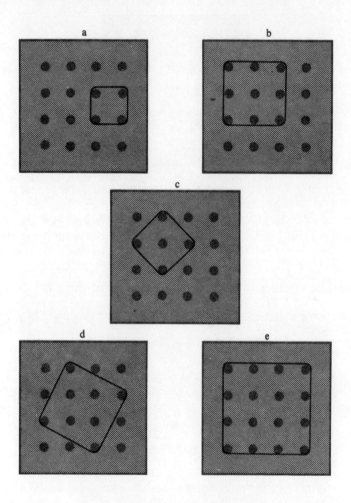

Now investigate the 25-pin board, and larger ones if you have them, in the same way. Keep a record of your results, as you will need them again later on.

Pegging squares

In this section you need to work with someone else (or even two or three others) using a piece of pegboard (hardboard with holes) and coloured pegs, one colour for each person. Instead of pegboard you could use a board or paper marked with squares and coloured counters to put in the squares.

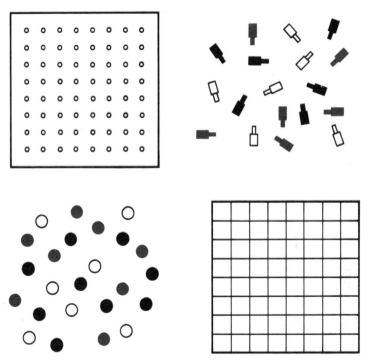

Each person in turn puts a peg anywhere in the board until one person has been able to make four of his pegs form the four corners of a square. Try this for yourself now. You will see that some squares are much harder to spot than others. When all the players are good at seeing squares then you will probably find that the board is almost covered with pegs before one player has won.

Can you see where a coloured peg should be put to complete a square on the board below?

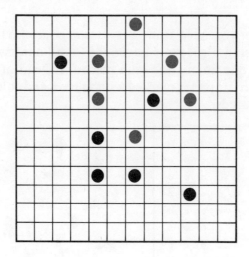

Now see if you find the fourth corner of a black square in the next diagram.

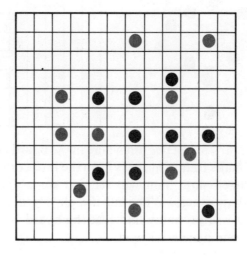

Growing squares

You have made pegboard squares from four pegs, one at each corner. Now make the *outlines* of some sequences of squares.

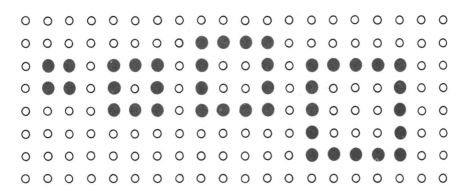

Draw up a table like this.

Pegs along one edge	2
Pegs in outline	4
Holes left inside	0

Why is the number of pegs in an outline square always a multiple of 4?

If a square needed 48 pegs for its outline how many would lie along each side? How many holes would be left inside?

When you have decided on your answers check them yourself on pegboard.

The number sequence for outline pegs was

                4   8   12   16   ...

You have seen that four pegs must be added to change from one outline to the next. How many pegs must also be *moved* to make the change in the following diagram?

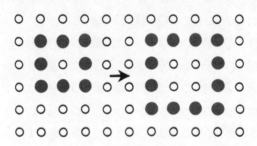

Investigate this for other changes from one outline to the next. Make a table and notice which special set of numbers turns up.

Pegs along edge	
Pegs added	
Pegs moved	
Pegs left alone	

Is it true that the sum of two consecutive odd numbers is *always* a multiple of four? Discuss this with someone else and explain how the pegboard helps you to illustrate whether the relationship is true or false.

You probably noticed that the holes left inside the outline squares also made a square formation and that the numbers of holes were:

0   1   4   9   16   25   ...

Illustrate this sequence by making filled squares on the pegboard.

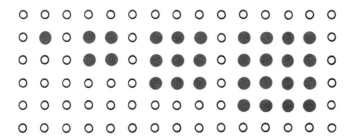

Study the number of pegs which have to be added to get from one square to the next and make a table.

Pegs along one edge	1	2
Total number of pegs	1	4
Additional pegs	3	

Find a quick way of working out the total number of pegs once you know the number along one edge of a filled square.

By now you will be able to make up your own pegboard investigations based on squares, so we leave you with just one small problem: make a filled-in square from exactly 13 pegs!

Square numbers

The numbers 1, 4, 9, 16... which you met in the last section are called SQUARE NUMBERS.

They can be obtained from $1 \times 1, 2 \times 2, 3 \times 3, 4 \times 4$... Sometimes the sequence is written $1^2, 2^2, 3^2, 4^2$... and read as *one squared*, *two squared*, *three squared*, *four squared* etc. To square a number sounds peculiar, but it just means to multiply it by itself.

Which square numbers come from the pegboard sequence in the diagram opposite?

What is the number sequence of pegs to be added each time?

Is it true that *every* even square number is a multiple of 4? Explain the reasons for your answer (you may find it useful to look at your work on outline squares again).

Build up and investigate the *odd* square numbers, 1, 9, 25... in the same way. Why are odd square numbers always 1 more than a multiple of 4?

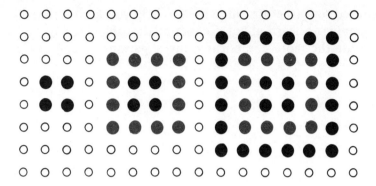

While you have worked on pegboard squares the sequence of odd numbers has turned up twice. There is a close connection between square numbers and odd numbers which you will see easily as you work out these sums:

1,   1 + 3,   1 + 3 + 5,   1 + 3 + 5 + 7,   1 + 3 + 5 + 7 + 9

How many odd numbers would you have to add together to get 64; 81; 100? Work out a general rule.

If you have already worked from the book in this series called *Triangles* this may have reminded you of the sequence of triangular numbers 1, 3, 6, 10 ... which comes from:

1,   1 + 2,   1 + 2 + 3,   1 + 2 + 3 + 4,   1 + 2 + 3 + 4 + 5

Use any apparatus you like, or just pencil and paper, to try to find an easy relationship between square and triangular numbers.

Coloured rods (the sort often used to help young children when they are beginning to learn about number relationships) can be very useful to illustrate the more complicated investigations which *you* are doing. If some rods are available build up this pattern of squares.

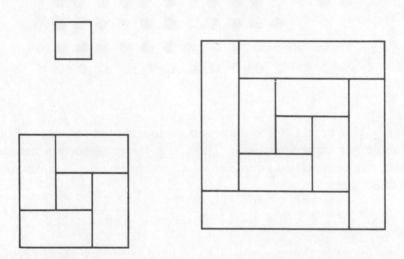

How can this pattern be used to show that the odd square numbers are always 1 more than a multiple of 4? (You met this statement on page 13, using pegs.)

Make a rod pattern for *even squares*, and work out a similar statement for them.

Use rod patterns to show that the difference between two successive odd squares is always a multiple of 8.

Investigate the differences between successive even squares.

Sums and differences

Here is the beginning of an addition table for two square numbers. Make one for yourself, going up as far as $20^2 + 20^2$, using a calculating machine, slide rule or table book if possible.

As you fill in the table, notice the number patterns, especially the patterns of differences between numbers, in the rows, columns and diagonals.

+	1^2	2^2	3^2	4^2	.	.	.
1^2	2	5	10	17	.	.	.
2^2	5	8	13	20	.	.	.
3^2	10	13	18	25	.	.	.
4^2	17	20	25	32	.	.	.
.	.	.	.	.	.	.	.
.	.	.	.	.	.	.	.
.	.	.	.	.	.	.	.

As you made your table did you notice the difference pattern 3, 5, 7, 9 ..., the odd numbers, again? You already know that this is the difference pattern for consecutive square numbers.

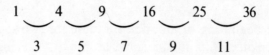

Why does this sequence turn up again in the 'addition-of-two-squares' table?

Make a window card the right size to pick out four adjacent numbers as you move it over your table.

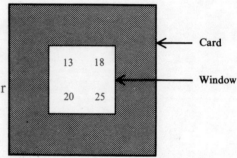

Try to find relationships which *always* exist between the four numbers showing through the window, no matter where you put it on the addition-of-squares table. Try to explain the reason for these relationships and discuss them with someone else who has been working on this section.

Another thing you may have noticed is the geometrical pattern of odd and even numbers. Mark all the even numbers in some way. Try to explain why this arrangement of odds and evens occurs.

Inside the pattern of even numbers is a pattern of multiples of 4. Mark this pattern in some way and explain why it happens.

Now look at the odd numbers and explain why they come up in their chequerboard arrangement. Make an order list of all the odd numbers in the table. **Are there any odd numbers which are not there and never would be no matter how long you went on listing?**

Divide each number in your list by 4 and look at the *remainder*. What do you notice? Can you explain why this happens? It may help if you look at these diagrams of the pegboard and coloured rods.

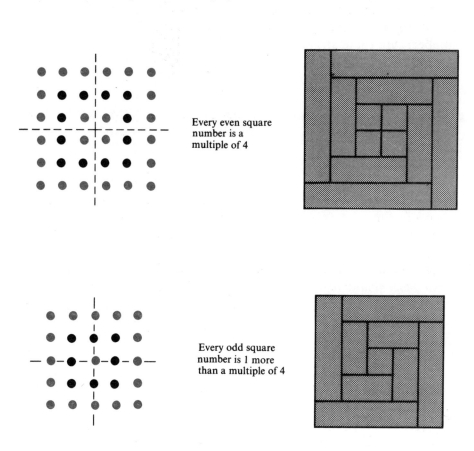

Every even square number is a multiple of 4

Every odd square number is 1 more than a multiple of 4

17

Here is a list of odd numbers arranged in two rows. Those in the top row are 1 more than a multiple of 4 and those in the bottom row are 1 less than a multiple of 4. Will any odd numbers be missed in this way of listing? In which row will the *odd square numbers* lie?

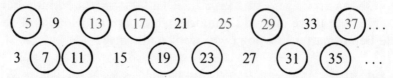

The numbers in red are the odd numbers which you have already listed from the sum-of-two-squares table. The numbers in the rings are PRIME NUMBERS. They have exactly two divisors (themselves and 1). Are there any *even* prime numbers? Is the number 1 itself a prime number? Are there any square primes?

The number of divisors of a number can tell you various things. For example the number of divisors can tell you whether a number is square or not. Investigate this.

The relationships between sets of numbers such as primes, squares, odds and evens have been studied and recorded for thousands of years. Many of them can be explained but there are still some which seem to be true but cannot be explained so far. One well known investigator in this subject of NUMBER THEORY was Fermat, a Frenchman (1601–1665). Amongst his many statements is one which says that *if a number is one more than a multiple of 4* **and** *if it is also a prime number then it is the sum of two square numbers – what is more it can only be done in one way.* Do the number sequences at the top of this page support Fermat's theory?

There are a lot of books about the history of mathematics and investigations into number relationships. Two useful ones for a school library are *Mathematics, the man made universe*, by S. K. Stein (published by W. H. Freeman and Co.) and *A short account of the history of mathematics*, by W. Rouse Ball (published by Dover Books, and Constable).

Some investigations can be based on the differences between square numbers. You have done some already. The results of one of them gave the odd numbers as the differences between successive squares.

$2^2 - 1^2 = 3$
$3^2 - 2^2 = 5$
$4^2 - 3^2 = 7$
$5^2 - 4^2 = 9$
$6^2 - 5^2 = 11$
. . . .

Try to find a link between the two numbers on the left and the odd number on the right. Does this link exist in every case? Check the prediction by using some bigger numbers like $50^2 - 49^2$.

Continue the investigation of differences using alternate square numbers.

Notice the pattern and look again for links between the numbers.
This one is not so easy!

$3^2 - 1^2 =$
$4^2 - 2^2 =$
$5^2 - 3^2 =$
. . . .

When the square numbers were consecutive you probably noticed a straightforward addition link. For example:

$$4^2 - 3^2 = 7 \qquad 4 + 3 = 7$$

You found that this idea always worked and it can be demonstrated easily with peg patterns.

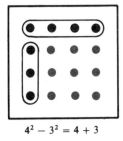

$4^2 - 3^2 = 4 + 3$

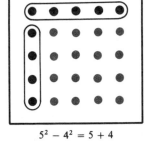

$5^2 - 4^2 = 5 + 4$

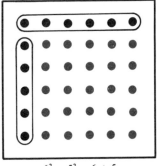

$6^2 - 5^2 = 6 + 5$

The difference sequence for alternate squares,

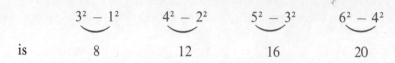

is   8   12   16   20

Addition of the two numbers to be squared gives:

        4   6   8   10

So just addition cannot be the way this time. However, maybe you have noticed that the numbers in the 8, 12, 16, 20 ... sequence are double those in the straightforward addition sequence 4, 6, 8, 10 ... This shows us that for alternate square numbers the link is *add and then multiply by 2*. The pegboard patterns show this well.

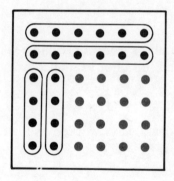

Check this formula for higher numbers. For example, applied to $25^2 - 23^2$ it would give

$$(25 + 23) \times 2 = 96$$

Is that really the difference between 25^2 and 23^2?

Now investigate square numbers which are three apart, like:

$$4^2 - 1^2 \quad 5^2 - 2^2 \quad 6^2 - 3^2$$

Look for a quick calculating method again. **Pegboard patterns give a strong clue.**

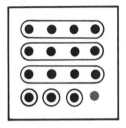
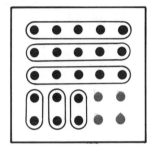

Have you got an idea yet for a general rule for any difference of two squares, no matter how far apart in the sequence of square numbers? If so check your method on a few different examples like:

$$10^2 - 5^2 \quad \text{and} \quad 16^2 - 4^2$$

The *difference-of-two-squares* on which you have been working is very well known and can be linked to ideas about area of squares and rectangles. This diagram shows how a rectangle can be made from the piece left when one square is removed from the corner of another.

Make a card model for yourself and see if you can find the connection between this and the general rule for the difference of two square numbers.

The areas of geometrical shapes are often used to demonstrate relationships between numbers. The one which is probably best known concerns the sums of two square numbers, which you have already studied. Sometimes the sum of two squares is *itself* a square number. Look through all your work, including the addition table for two squares, and list any examples which you find. Use your knowledge of the difference between two squares to try to find other examples. You may be able to add a completely new one to a class collection.

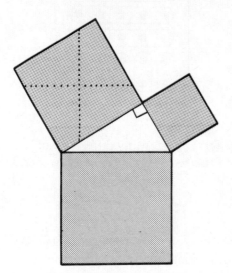

This diagram, which you will find in a great many books, shows squares on the three sides of a right-angled triangle. The dotted lines show how to cut one square so that the two smaller squares can be fitted into the larger. Make a cardboard model yourself based on any right-angled triangle you like and show how the pieces fit. Make a model based on a triangle which is not right-angled.

Take some of the relationships you found, such as $3^2 + 4^2 = 5^2$ and $5^2 + 12^2 = 13^2$, and cut out card squares to represent them. What is special about the triangles you form by putting the related squares corner to corner?

Three square numbers linked in the form $a^2 + b^2 = c^2$ are called Pythagorean and Pythagoras's Theorem, which concerns the relationship between the squares of the sides of a right-angled triangle, is one of the most useful applications of number relationship which has ever been found.

Square pyramids

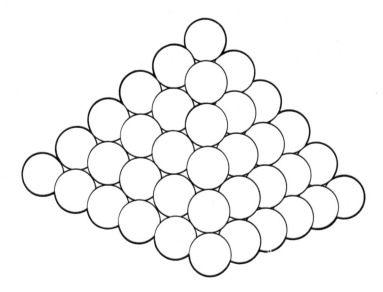

How many spheres do you think were used to make the square-based pyramid in the diagram? Perhaps you have seen fruit piled up like this in a greengrocer's shop.

Collect some equal-sized marbles, table tennis balls, beads, polystyrene spheres or other spherical objects and build some pyramids yourself. (It may help if you press the base layer into some flattened plasticene to stop it slipping.)

Make a list of the number of spheres in each layer. **What do you notice about the numbers in your list?**

How many spheres would you need altogether for a four-layer square pyramid? How many would you need for a five-layer pyramid? Make a table like this:

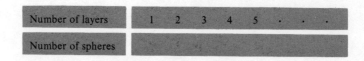

Number of layers	1	2	3	4	5	.	.	.
Number of spheres								

Now work out the number of spheres in the pyramid in the diagram on page 23. Were you far out in the guess you made when you started the chapter?

You can make pyramids by building them up in a stiff cardboard shell like the one shown here.

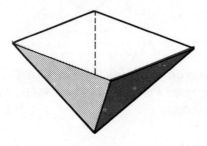

If you have worked through *Circles* you may remember that a chemist can use a triangular tray to count pills quickly. You could use a pyramid shell in a similar way to count spheres. Make a shell and tip some identical spheres into it. How can you tell how many layers there are? Work out how many spheres you have tipped into the shell.

Why is a pyramid shell better than a triangular tray for counting spheres?

These two pyramids are built of cubes.

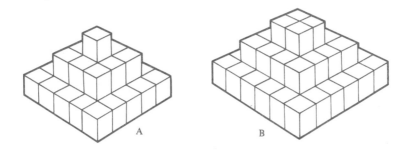

How many cubes were needed for each pyramid? Why are these two numbers different from the numbers for pyramids made of spheres?

Look at the numbers in column (2) of this table. What do they represent? Copy the table and fill in the rest of columns (2) and (3).

(1)	(2) Type A pyramid	(3) Type B pyramid	(4)
1 layer	1	4	3
2 layers	10	20	10
3 layers	35	56	21
4 layers			
5 layers			
6 layers			

How do the numbers in column (4) come from (2) and (3)? Complete column (4) and try to find a pattern in the way the numbers grow. They are from a special set, called triangular numbers, which you will know well if you have worked from the booklet called *Triangles*. Can you explain why they turn up here?

25

Counting squares

How many squares are there in this diagram?

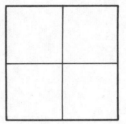

It is easy to see four squares but did you see a *fifth* one?

How many squares can you find in this next diagram?

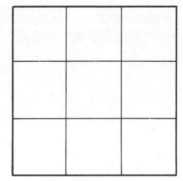

You probably found the nine smallest ones

and the largest one

Did you get these?

How many of them are there?

How many of the 4-unit size are there in the 16-unit square below?

27

Did you find the four at the corners –

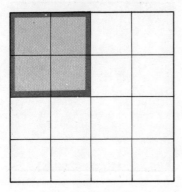

– the four from the middle of the sides –

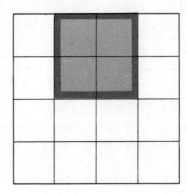

– and the one on its own in the centre –

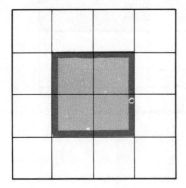

– making nine altogether?

Now try to analyse *all* the squares in a 25-unit diagram.

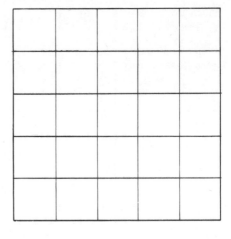

Here is a square which can be made, first of all, on a 16-pin board.

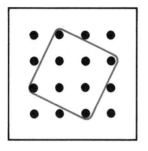

See how many of these you can make on a 16, 25, 36 and 49 pin board. Make a table of results and check them with someone else.

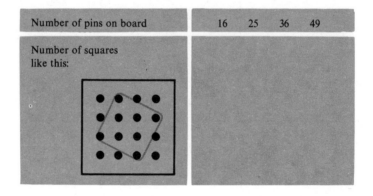

Number of pins on board	16	25	36	49
Number of squares like this:				

How are the numbers you have found related to the square numbers 1, 4, 9 ...?

There are still more sizes of squares you can count on the nailboards. As there is a lot of work in this, perhaps a group could work together on boards of different sizes. When you have done it and made tables of the results you will be able to discover a lot more patterns based on the **square numbers.** As a last problem try to show how you could *work out*, without counting them all separately, the total number of squares which could be made on any size nailboard.

You have already made **squares on a nailboard**. Turn back to page 6 to remind yourself of the five different sizes that you made on the 16-pin board. *How many* of each size square can you make? Try it on a board or on dotted paper. Then try to find the number of squares of each size on a 25-pin board.

Here is one possible square on a board of 9 or more pins.

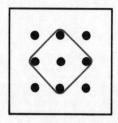

How many of these can you make on a 16-pin board? A 25-pin board? A 36-pin board?

Make a table

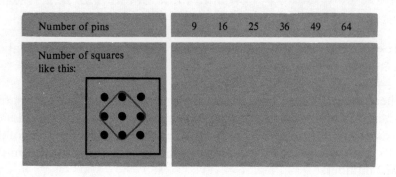

Number of pins	9	16	25	36	49	64
Number of squares like this:						

What do you notice about the numbers you have put in the table?

31

Original diagram	Unit size of square						Total
	1 unit	4 units	9 units	16 units	25 units	36 units	
☐	1						1
(2×2)	4	1					5
(3×3)	9	4	1				14
(4×4)							
(5×5)							

Did you find five different sizes? There was the 1-unit size, the 4-unit size, the 9-unit size, the 16-unit size and the 25-unit size.

Compare your answers with someone else's and make a table to show the number of squares of each size.

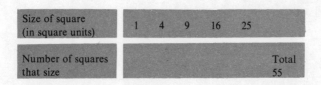

Size of square (in square units)	1	4	9	16	25	
Number of squares that size						Total 55

Is there any pattern in the numbers in the table?
Which special numbers have appeared?
Try to explain *why* they appeared.

You can now take this investigation further. Complete, and if possible continue, the table on the next page. It shows the number of squares of each size in the types of diagram you have been using.

When you have finished the table look at the numbers in the 'total' column. Where did they turn up before in this book? Can you *predict* the result for a 64-unit diagram?

When you began this book you tried to count the total number of squares of different sizes on a chessboard. Now check your answer without counting all the squares!